KINGS CANYON NATIONAL PARK
ACTIVITY BOOK

PUZZLES, MAZES, GAMES, AND MORE ABOUT
KINGS CANYON NATIONAL PARK

NATIONAL PARKS ACTIVITIES SERIES

KINGS CANYON NATIONAL PARK ACTIVITY BOOK

Copyright 2021
Published by Little Bison Press

The author acknowledges that the land that is now Kings Canyon National Park are the traditional lands of the Western Mono (Monache), the Foothills Yokuts, and the Tubatulabal People.

LITTLE BISON
Press

For more free national parks activities, visit
www.littlebisonpress.com

About Kings Canyon National Park

Kings Canyon National Park is located in the state of California. People from all over the world come to visit Kings Canyon to see the giant sequoias, some of the biggest trees in the world. Kings Canyon is located adjacent to Sequoia National Park.

Kings Canyon National Park was once called "a rival to Yosemite" by John Muir. This park is famous for its distinctive granite features and is a popular destination for hiking, camping, and horseback riding.

Fires are an important part of the landscape, both natural fires, and fires from prescribed burns. Some fires are actually beneficial to giant sequoias, because flames clear the ground of the forest floor. Fire also opens sequoia cones, causing seeds to fall down on the ground. In addition, fires remove ground vegetation that competes with the seedlings for moisture, nutrients, and sunshine.

Kings Canyon National Park is famous for:
- groves of the big trees
- the deep Kings Canyon and Kings River
- rock climbing

Hey, I'm Parker!

I'm the only snail in history to visit every National Park in the United States! Come join me on my adventures in Kings Canyon National Park.

hroughout this book, we will learn about the history of the park, the animals and plants that live here, and things to do if you ever visit in person. This book is also full of games and activities!

Last but not least, I am hidden 9 times on different pages. See how many times you can find me. This page doesn't count!

Kings Canyon Bingo

Let's play bingo! Cross off each box you are able to during your visit to the national park. Try to get a bingo down, across, or diagonally. If you can't visit the park, use the bingo board to plan your perfect trip.

Pick out some activities you would want to do during your visit. What would you do first? How long would you spend there? What animals would you try to see?

SPOT A MAMMAL SMALLER THAN YOU	HUG A GIANT SEQUOIA	IDENTIFY A TREE	TAKE A PICTURE AT AN OVERLOOK	WATCH A MOVIE AT THE VISITORS CENTER
GO FOR A HIKE	LEARN ABOUT THE INDIGENOUS PEOPLE WHO LIVE IN THIS AREA	WITNESS A SUNRISE OR SUNSET	OBSERVE THE NIGHT SKIES	VISIT CRYSTAL CAVE
HEAR A BIRD CALL	SPOT A RUSHING RIVER	FREE SPACE	LEARN ABOUT THE IMPORTANCE OF FIRES	VISIT A RANGER STATION
PICK UP 10 PIECES OF TRASH	GO CAMPING	SEE A MULE DEER	VISIT GRANT GROVE	SPOT A BIRD OF PREY
LEARN ABOUT THE GEOLOGY OF THE SIERRAS	SEE SOMEONE RIDING A HORSE	HAVE A PICNIC	SPOT SOME ANIMAL TRACKS	PARTICIPATE IN A RANGER-LED ACTIVITY

The National Park Logo

The National Park System has over 400 units in the US. Just like Kings Canyon National Park, each location is unique or special in some way. The areas include other national parks, historic sites, monuments, seashores, and other recreation areas.

Each element of the National Park emblem represents something that the National Park Service protects. Fill in each blank below to show what each symbol represents.

```
WORD BANK:

MOUNTAINS, ARROWHEAD, BISON,
SEQUOIA TREE, WATER
```

This represents all plants: _____

This represents all animals: _____

This represents the landscapes: _____

This represents the waters protected by the park service: _____

This represents the historical and archeological values: _____

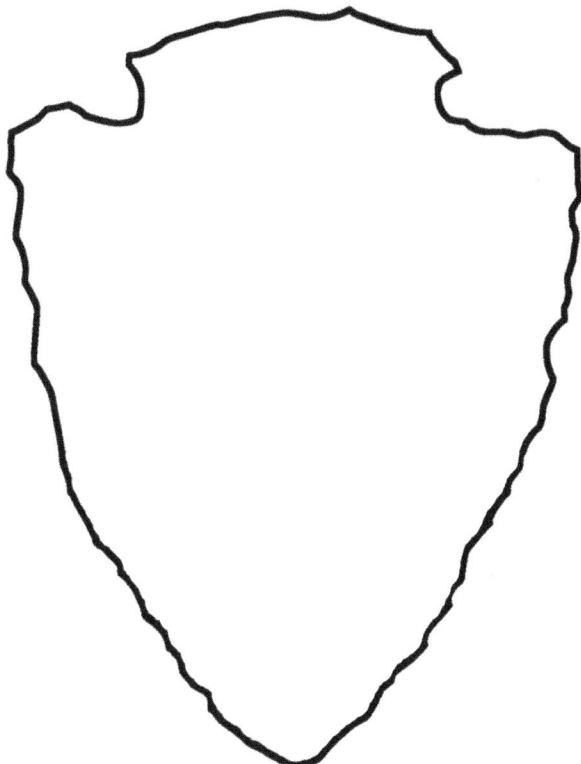

Now it's your turn! Pretend you are designing a new national park. Add elements to the design that represent the things your park protects.

What is the name of your park?

Describe why you included the symbols that you chose. What do they mean?

Things to Do Jumble

Unscramble the letters to uncover activities you can do while in Kings Canyon National Park. Hint: each one ends in -ing.

1. MRLIKCCOB
   ☐☐☐☐■☐☐☐☐☐ ING

2. KHI
   ☐☐☐ ING

3. IRDB
   ☐☐☐☐ ING

4. MACP
   ☐☐☐☐ ING

5. KINICPC
   ☐☐☐☐☐☐☐ ING

6. EISSTEHG
   ☐☐☐☐☐☐☐☐ ING

7. SHOSNOWE
   ☐☐☐☐☐☐☐☐ ING

Word Bank

birding
reading
camping
snowshoeing
rock climbing
hiking
hunting
singing
yelling
sightseeing
picnicking

Photobook

Draw some pictures of
things you saw in the park.

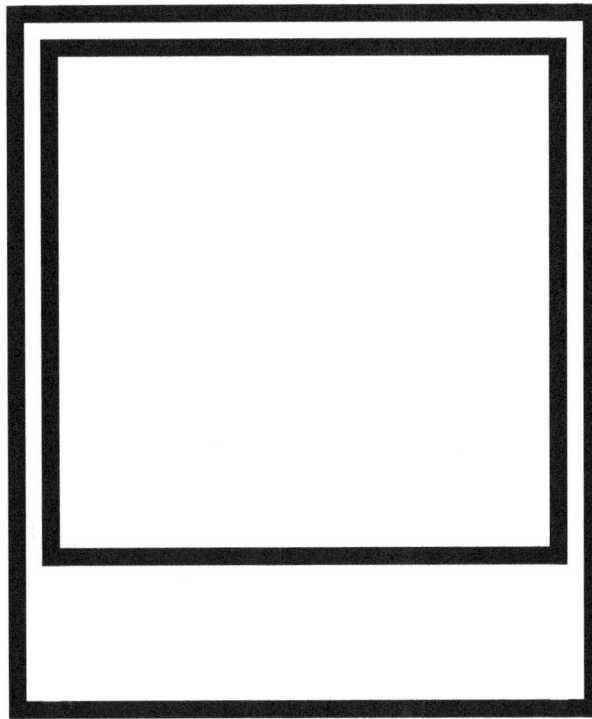

Go Birdwatching at Zumwalt Meadow

start here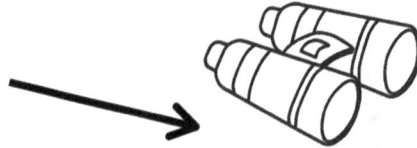

Camping Packing List

What should you take with you when you go camping? Pretend you are in charge of your family camping trip. Make a list of what you would need to be safe and comfortable on an overnight excursion. Some considerations are listed on the side.

1.
2.
3.
4.
5.
6.
7.
8.
9.
10.
11.
12.
13.
14.
15.
16.

- What will you eat at every meal?

- What will the weather be like?

- Where will you sleep?

- What will you do during your free time?

- How luxurious do you want camp to be?

- How will you cook?

- How will you see at night?

- How will you dispose of trash?

- What might you need in case of emergencies?

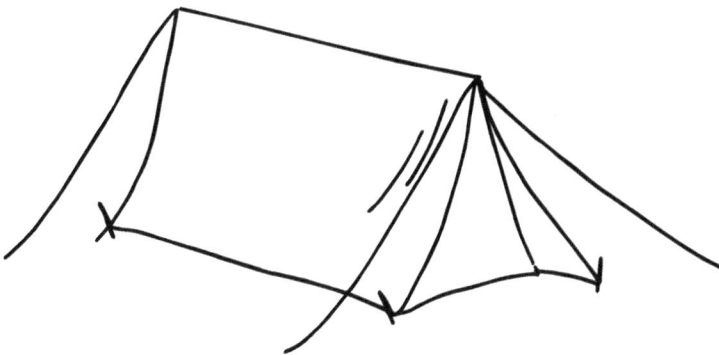

Kings Canyon National Park

Date: _____

Season: _____

Who I went with: _____

Which entrance: _____

How was your experience? Write a few sentences about your trip. Where did you stay? What did you do? What was your favorite activity? If you haven't visited the park yet, write a paragraph pretending that you did.

_____

_____

_____

_____

_____

STAMPS

Many national parks and monuments have cancellation stamps for visitors to use. These rubber stamps record the date and location that you visited. Many people collect the markings as a free souvenir. Check with a ranger to see where you can find a stamp during your visit. If you aren't able to find one, you can draw your own.

Where is the Park?

Kings Canyon National Park is in the western United States. It is located in California. There are a total of 9 national parks in California, more than any other state!

California

Look at the shape of California. Can you find it on the map? If you are from the US, can you find your home state? Color California red. Put a star on the map where you live.

Connect the Dots #1

Connect the dots to figure out what this tiny critter is. There are five types of these that live in Kings Canyon National Park.

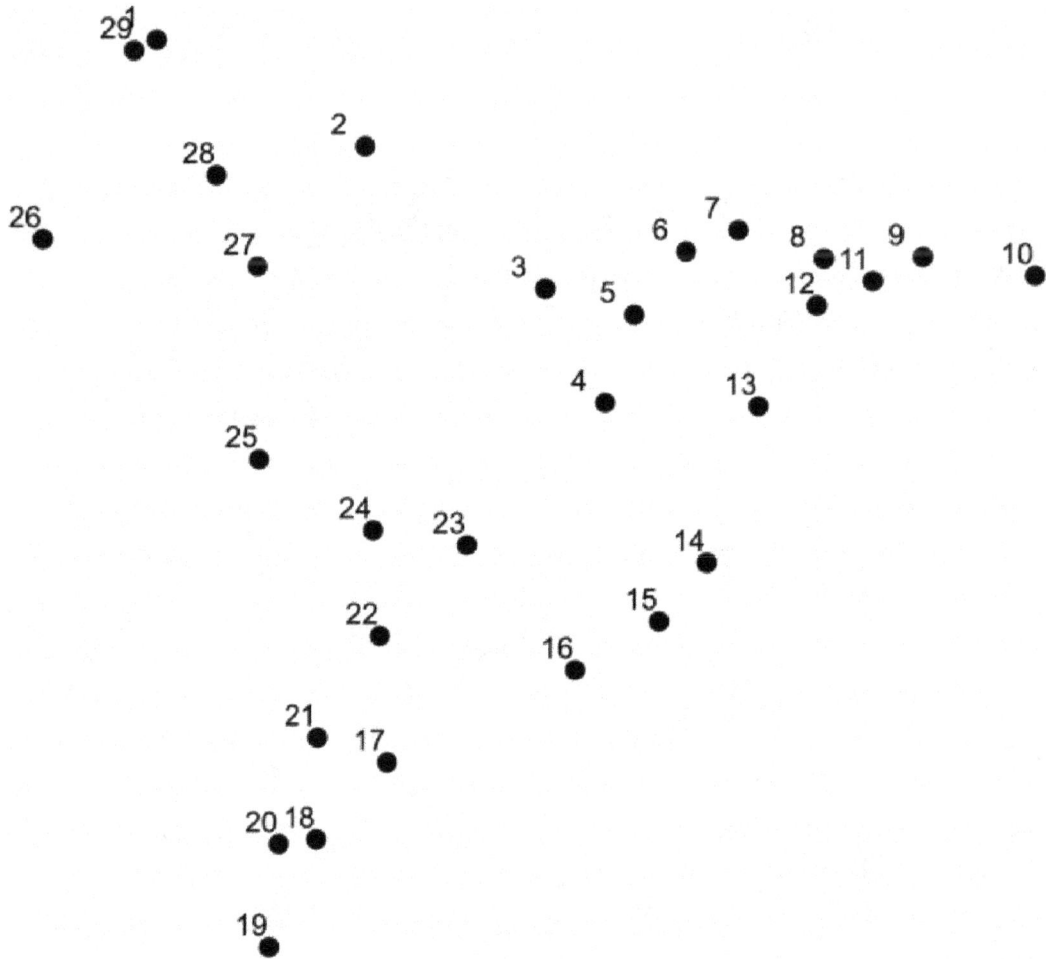

29 1
2
28
26
27
7
6 8 9
3 11 10
5 12
4 13
25
24 23
14
15
22
16
21
17
20 18
19

Their heart rate can reach as high as 1,260 beats per minute and a breathing rate of 250 breaths per minute. Have you ever measured your breathing rate? Ask a friend or family member to set a timer for 60 seconds. Once they say "go," try to breathe normally. Count each breath until they say "stop." How do your breaths per minute compare to hummingbirds?

_____

_____

Pikas are related to rabbits, but they have distinctly rounded ears. They can be hard to spot but you might hear them. Listen for their short, high-pitched call.

Mountain lions, also known as cougars, are rarely seen by humans. They favor remote, forested places in the park.

Who Lives Here?

Below are 8 plants and animals that live in the park.
Use the word bank to fill in the clues below.

WORD BANK: GOPHER SNAKE, QUAIL, POSSUM, MARMOT, WESTERN TOAD, PIKA, BEAVER, WILD TURKEY

☐ ☐ ☐ S ☐ ☐

☐ ☐ ☐ ☐ E ☐ ☐ ■ ☐ ☐ ☐ ☐

Q ☐ ☐ ☐ ☐

☐ ☐ ☐ ☐ ■ ☐ U ☐ ☐ ☐ ☐

☐ ☐ ☐ ☐ O ☐

☐ I ☐ ☐

☐ ☐ A ☐ ☐ ☐

☐ ☐ ☐ ☐ ☐ ☐ ■ S ☐ ☐ ☐ ☐

Porcupines are well known for their defense mechanism, their quills! If attacked, these quills easily detach from the porcupine's back to pierce potential predators.

Beavers are the largest North American rodent.

Common Names
vs.
Scientific Names

A common name of an organism is a name that is based on everyday language. You have heard the common names of plants, animals, and other living things on tv, in books, and at school. Common names can also be referred to as "English" names, popular names, or farmer's names. Common names can vary from place to place. The word for a particular tree may be one thing, but that same tree has a different name in another country. Common names can even vary from region to region, even in the same country.

Scientific names, or Latin names, are given to organisms to make it possible to have uniform names for the same species. Scientific names are in Latin. You may have heard plants or animals referred to by their scientific name or parts of their scientific names. Latin names are also called "binomial nomenclature," which refers to a two-part naming system. The first part of the name – the generic name – refers to the genus to which the species belongs. The second part of the name, the specific name, identifies the species. For example, Tyrannosaurus rex is an example of a widely known scientific name.

American Black Bear

Ursus americanus

COMMON NAME

Brewer's Blackbird

Euphagus cyanocephalus

LATIN NAME = GENUS + SPECIES

Brewer's Blackbird = Euphagus cyanocephalus

Black Bear = Ursus americanus

Find the Match!
Common Names and Latin Names

Match the common name to the scientific name for each animal. The first one is done for you. Use clues on the page before and after this one to complete the matches.

Brewer's Blackbird   Haliaeetus leucocephalus

Giant Sequoia   Ursus americanus

Corn Lily   Lagopus leucura

American Black Bear   Ochotona princeps

Great Horned Owl   Sequoiadendron giganteum

Bald Eagle   Charina bottae

Ptarmigan   Bubo virginianus

Pika   Euphagus cyanocephalus

Rubber Boa   Veratrum californicum

Bald Eagle
Haliaeetus leucocephalus

17

Ptarmigan
Lagopus leucura

Bald Eagle
Haliaeetus leucocephalus

Great Horned Owl
Bubo virginianus

Some plants and animals that live at Kings Canyon

Giant Sequoia
Sequoiadendron giganteum

Pika
Ochotona princeps

Rubber Boa
Charina bottae

Wildlife Wisdom

The national park is home to many different kinds of animals. Seeing wildlife can be an exciting part of visiting the national park but it is important to remember that these animals are wild. They need plenty of space and a healthy habitat where they can find their own food. Part of this is not allowing animals to eat any human food. This is their home and we are the visitors. We need to be respectful of the wildlife in the park.

Directions: Circle the highlighted words that best complete the following sentences.

If an animal changes its behavior because of your presence, you are:
A) too close
B) funny looking
C) dehydrated and should drink more water

The best thing we can do to help wild animals survive is:
A) make them pets
B) protect their habitat
C) knit them winter sweaters

In a national park, it is okay to share your food with wild animals:
A) never
B) always
C) sometimes

When you're hiking in an area where there are bears, you should warn bears that you are entering their space by:
A) hiking quietly
B) making noise
C) wearing bright colors

At night, park rangers care for the animals by:
A) putting them back into their cages
B) tucking them into bed
C) leaving them alone

If you see an abandoned bird's nest, it is best to:
A) pet the baby birds
B) leave it alone
C) crunch the empty eggshells

Bears look under logs in hopes of finding:
A) granola bars
B) insects
C) peanuts to eat

The place where an animal lives is called its:
A) condo
B) habitat
C) crib

The Ten Essentials

Careful preparation and knowledge are key to a successful adventure into Kings Canyon backcountry.

The ten essentials are a list of things that are important to have when you go for longer hikes. If you go on a hike in the <u>backcountry</u>, it is especially important that you have everything you need in case of an emergency. If you get lost or something unforeseen happens, it is good to be prepared to survive until help finds you.

The ten essentials list was developed in the 1930s by an outdoors group called the Mountaineers. Over time and technological advancements, this list has evolved. Can you identify all the things on the current list? Circle each of the "essentials" and cross out everything that doesn't make the cut.

fire: matches, lighter, tinder, and/or stove	a pint of milk	extra money	headlamp, plus extra batteries	extra clothes
extra water	a dog	Polaroid camera	bug net	lightweight games, like a deck of cards
extra food	a roll of duct tape	shelter	sun protection, such as sunglasses, sun-protective clothes, and sunscreen	knife, plus a gear repair kit
a mirror	navigation: map, compass, altimeter, GPS device, or satellite messenger	first aid kit	extra flip-flops	entertainment, such as video games or books

Backcountry - a remote, undeveloped rural area.

Draw A Meal

Imagine you've been adventuring all day and now it's time to go back to camp for the night. You are hungry!

Draw the meal that you will cook over the campfire.

Connect the Dots #2

This animal lives in almost every state in the US, including the national park. They are nocturnal, more active at night, and sleep during the day. They are omnivorous eaters, meaning they eat both plants and animals.

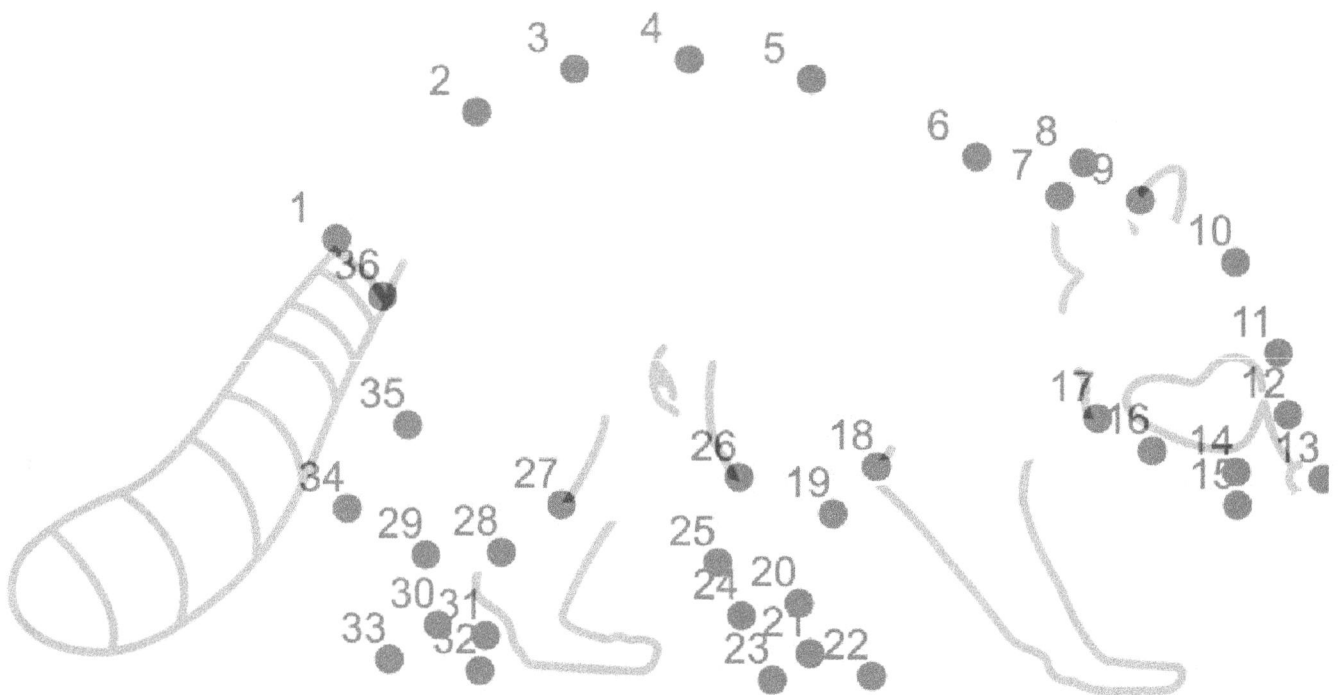

Are you an omnivore like a raccoon? An herbivore only eats plant foods. A carnivore only eats meat. An omnivore eats both. What type of eater are you? Write down some of your favorite foods to back up your answer.

_____

_____

LISTEN CAREFULLY

Visitors to Kings Canyon National Park may hear different noises from those they hear at home. Try this activity to experience this for yourself!

First, find a place outside where it is comfortable to sit or stand for a few minutes. You can do this by yourself or with a friend or family member. Once you have a good spot, close your eyes and listen. Be quiet for one minute and pay attention to what you are hearing. List some of the sounds you have heard in one of the two boxes below:

NATURAL SOUNDS
MADE BY ANIMALS, TREES OR PLANTS, THE WIND, ETC

HUMAN-MADE SOUNDS
MADE BY PEOPLE, MACHINES, ETC

ONCE YOU ARE BACK AT HOME, TRY REPEATING YOUR EXPERIMENT:

NATURAL SOUNDS
MADE BY ANIMALS, TREES OR PLANTS, THE WIND, ETC

HUMAN-MADE SOUNDS
MADE BY PEOPLE, MACHINES, ETC

WHERE DID YOU HEAR MORE NATURAL SOUNDS? _____

WHERE DID YOU HEAR MORE HUMAN SOUNDS? _____

Sound Exploration

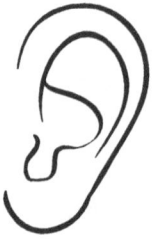

Spend a minute or two listening to all of the sounds around you.
Draw your favorite sound.

How did this sound make you feel?

_____

_____

What did you think when you heard this sound?

_____

_____

Kings Canyon Word Search

Words may be horizontal, vertical, diagonal,
or they might even be backwards!

1. giant sequoia
2. Fresno
3. trees
4. meadow
5. coyote
6. streams
7. California
8. cedar grove
9. pinecones
10. Panoramic Point
11. marble
12. Crystal Cave
13. pikas
14. marmot
15. skunk
16. forest
17. black bear

```
G W S L S P I R E F O R E S T
H I A S K I L S T R E A M S J
M E A D O W O S C E L B A P B
S P I N E C O N E S R L U A C
C E A D T A B L O N I U J N L
A T R E E S C O Y O T E A O I
L E E T E H E K I N K R I R N
I L V A M B I Q G W N E K A G
F E A S G L L O U E I D Y M M
O C C E D A R G R O V E O I A
R T L H C C I N O O I E M C N
N R A I K K E I S M O A I P E
I I T S H B I R E I R A L O W
A C S O L E V E S B S A K I P
N I Y K K A I N L R A L H N A
X T R F U R E E L Z E S Q T L
H Y C R O N L E C T R I C E E
F L O Y D N K D N T O M R A M
```

Find the Match!
What are Baby Animals Called?

Match the animal to its baby. The first one is done for you.

Elk   eaglet

Bald Eagle   calf

Little Brown Bat   snakelets

Striped Skunk   pup

Great Horned Owl   owlet

Western Toad   kit

Mountain Lion   tadpole

Garter snake   kitten

Protecting the Park

When you visit national parks, it is important to leave the park the way you found it. Did you know that the national parks get hundreds of millions of visitors every year? We can only protect national parks for future visitors to enjoy if everyone does their part. The choices that each visitor makes when visiting the park have a big impact all together.

Read each line below. Write a sentence or draw a picture to show the impacts these changes would make on the park.

What would happen if every visitor fed the wild animals?

What would happen if every visitor picked a flower?

What would happen if every visitor took home a few rocks?

What would happen if every visitor wrote or carved their name on the rocks or trees?

The Perfect Picnic Spot

Fill in the blanks on this page without looking at the full story. Once you have each line filled out, use the words you've chosen to complete the story on the next page.

EMOTION _____

FOOD _____

SOMETHING SWEET _____

STORE _____

MODE OF TRANSPORTATION _____

NOUN _____

SOMETHING ALIVE _____

SAUCE _____

PLURAL VEGETABLES _____

ADJECTIVE _____

PLURAL BODY PART _____

ANIMAL _____

PLURAL FRUIT _____

PLACE _____

SOMETHING TALL _____

COLOR _____

ADJECTIVE _____

NOUN _____

A DIFFERENT ANIMAL _____

FAMILY MEMBER #1 _____

FAMILY MEMBER #2 _____

VERB THAT ENDS IN -ING _____

A DIFFERENT FOOD _____

The Perfect Picnic Spot

Use the words from the previous page to complete a silly story.

When my family suggested having our lunch at the Grizzly Falls campground, I

was _ _ _ _ _ _ _ _. I love eating my _ _ _ _ _ _ outside! I knew we had picked up a
      EMOTION   FOOD

box of _ _ _ _ _ _ from the _ _ _ _ _ _ _ _ for after lunch, my favorite. We drove up
    SOMETHING SWEET   STORE

to the area and I jumped out of the _ _ _ _ _ _ _ _. "I will find the perfect spot for
                   MODE OF TRANSPORTATION

a picnic!" I grabbed a _ _ _ _ _ _ for us to sit on, and I ran off. I passed a picnic
          NOUN

table, but it was covered with _ _ _ _ _ _ _ _ so we couldn't sit there. The next
         SOMETHING ALIVE

picnic table looked okay, but there were smears of _ _ _ _ _ _ _ and pieces of
                  SAUCE

_ _ _ _ _ _ _ _ everywhere. The people that were there before must have been
PLURAL VEGETABLES

_ _ _ _ _ _! I gritted my _ _ _ _ _ _ _ together and kept walking down the path,
ADJECTIVE   PLURAL BODY PART

determined to find the perfect spot. I wanted a table with a good view of the

waterfall. Why was this so hard? If we were lucky, I might even get to see

_ _ _ _ _ _ eating some _ _ _ _ _ _ on the cliffside. They don't have those in
ANIMAL   PLURAL FRUIT

_ _ _ _ _ _ _ where I am from. I walked down a little hill and there it was, the
PLACE

perfect spot! The trees towered overhead and looked as tall as _ _ _ _ _ _ _ _. The
                  SOMETHING TALL

patch of grass was a beautiful _ _ _ _ _ _ _ color. The _ _ _ _ _ _ flowers were
        COLOR   ADJECTIVE

growing on the side of a _ _ _ _ _ _ _. I looked across the waterfall and even saw a
        NOUN

_ _ _ _ _ _ _ _ on the edge of a rock. I looked back to see my _ _ _ _ _ _ _ _ _ and
DIFFERENT ANIMAL   FAMILY MEMBER #1

_ _ _ _ _ _ _ _ _ _ _ _ _ _ _ _ _ a picnic basket. "I hope you brought plenty of
FAMILY MEMBER #2   VERB THAT ENDS IN ING

_ _ _ _ _ _ _ _, I'm starving!"
A DIFFERENT FOOD

29

Hike to see the Giant Sequoias

start here →

DID YOU KNOW?

The General Grant tree in Kings Canyon National Park is the second largest tree in the world by volume.

The Biggest Trees
Word Search

Giant Sequoias are some of the biggest trees in the world. Many of the very largest trees are named and closely monitored. These are some of the biggest ones in Kings Canyon and Sequoia National Park. Can you find them?

1. General Sherman
2. General Grant
3. President
4. Lincoln
5. Stagg
6. Boole
7. Genesis
8. Franklin
9. Monroe
10. Column
11. Euclid
12. Pershing
13. Diamond
14. Adams
15. Nelder
16. Hart

```
G D E S C A N Y O N D E D W C
E E D P M I F R A N K L I N H
N V N R K I T E A W A L A O A
E E U E H A R T U Y U T M M T
S N N S R D Y P L C Y R O K R
I P D I Y A R E C T L E N O E
S O S D P M L R R H L I D A E
A R B E M S I S D I L S D N G
L T H N G I L H H N U D E C E
S I S T A G G I K E U G R O R
E S N U A E I N N L R B N L N
Q H N C K N O G S D S M T U E
U J O S O E I N Z E I O A M C
O Y G E L L V E I R D N V N O
I W E L D A N A D O A R H E M
A T G E N E R E N L B O O L E
U A E E S A E N N O A E V E B
C G E N E R A L G R A N T O N
```

Leave No Trace Quiz

Leave No Trace is a concept that helps people make decisions during outdoor recreation that protects the environment. There are seven principles that guide us when we spend time outdoors, whether you are in a national park or not. Are you an expert in Leave No Trace? Take this quiz and find out!

1. How can you plan ahead and prepare to ensure you have the best experience you can in the national park?
   a. Make sure you stop by the ranger station for a map and to ask about current conditions.
   b. Just wing it! You will know the best trail when you see it.
   c. Stick to your plan, even if conditions change. You traveled a long way to get here, and you should stick to your plan.
2. What is an example of traveling on a durable surface?
   a. Walking only on the designated path.
   b. Walking on the grass that borders the trail if the trail is very muddy.
   c. Taking a shortcut if you can find one because it means you will be walking less.
3. Why should you dispose of waste properly?
   a. You don't need to. Park rangers love to pick up the trash you leave behind.
   b. You should actually leave your leftovers behind, because animals will eat them. It is important to make sure they aren't hungry.
   c. So that other peoples' experiences of the park are not impacted by you leaving your waste behind.
4. How can you best follow the concept "leave what you find?"
   a. Take only a small rock or leaf to remember your trip.
   b. Take pictures, but leave any physical items where they are.
   c. Leave everything you find, unless it may be rare like an arrowhead, then it is okay to take.
5. What is not a good example of minimizing campfire impacts?
   a. Only having a campfire in a pre-existing campfire ring.
   b. Checking in with current conditions when you consider making a campfire.
   c. Building a new campfire ring in a location that has a better view.
6. What is a poor example of respecting wildlife?
   a. Building squirrel houses out of rocks so the squirrels have a place to live.
   b. Stay far away from wildlife and give them plenty of space.
   c. Reminding your grown-ups not to drive too fast in animal habitats while visiting the park.
7. How can you show consideration of other visitors?
   a. Play music on your speaker so other people at the campground can enjoy it.
   b. Wear headphones on the trail if you choose to listen to music.
   c. Make sure to yell "Hello!" to every animal you see at top volume.

Park Poetry

America's parks inspire art of all kinds. Painters, sculptors, photographers, writers, and artists of all mediums have taken inspiration from natural beauty. They have turned their inspiration into great works.

Use this space to write your own poem about the park. Think about what you have experienced or seen. Use descriptive language to create an acrostic poem. This type of poem has the first letter of each line spell out another word. Create an acrostic that spells out the word "Forest."

F _____

O _____

R _____

E _____

S _____

T _____

F orests so tall
O ver the mountains
R ushing river
E ach tree so big
S equoias
T ell everyone how cool

F resh and early
O nly us on the road
R eady to sled
E ach mitten on
S now everywhere!
T reats to warm up

Build a Bird Nest

Different birds build different kinds of nests where they can lay their eggs and raise their babies.

Draw a nest and some baby birds that you might find in Kings Canyon National Park.

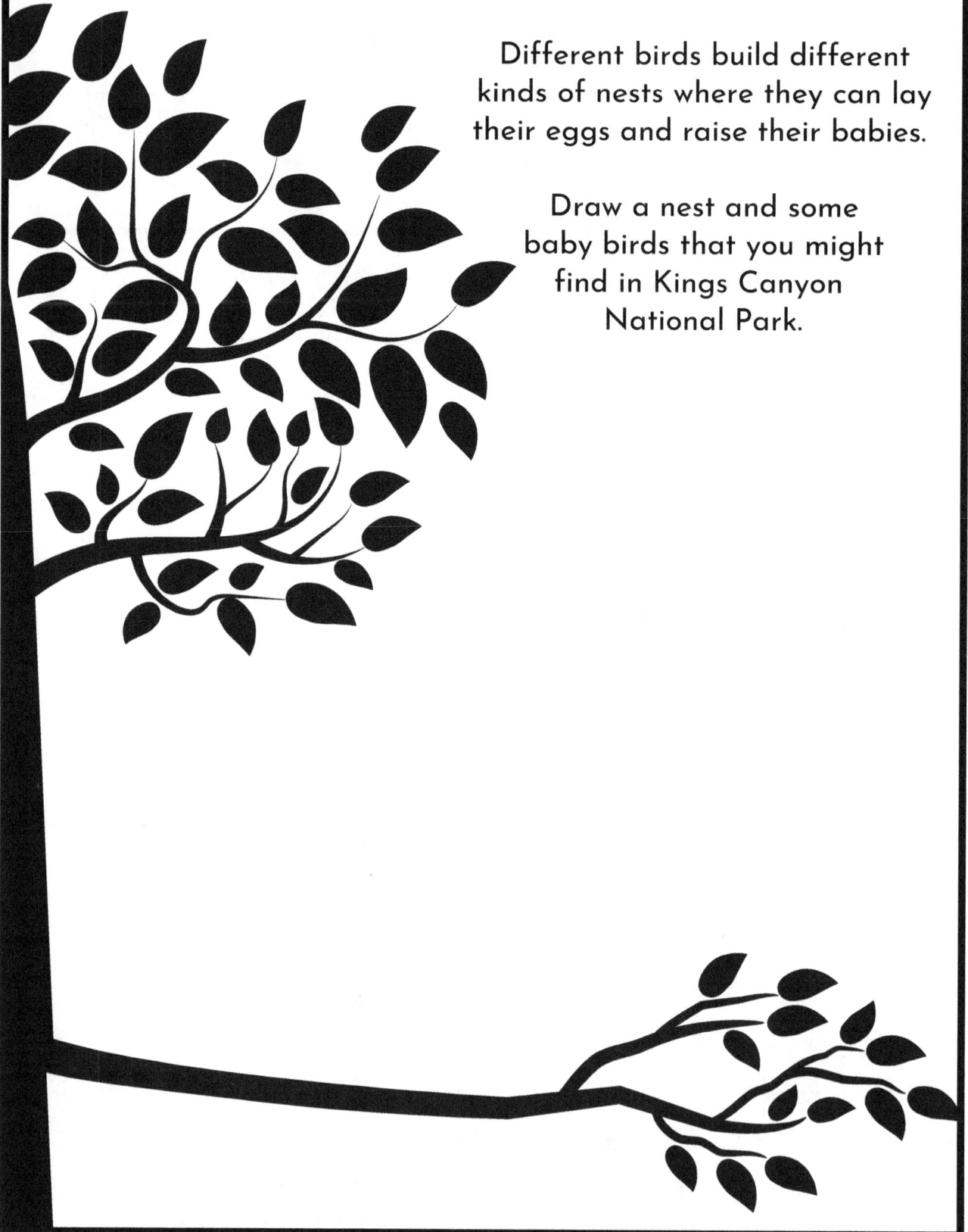

Catch a Fish in the Kings River

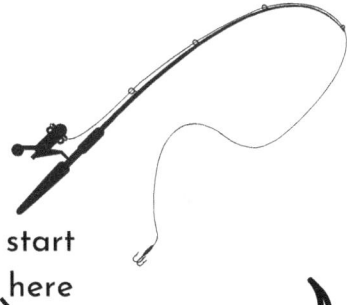

start here

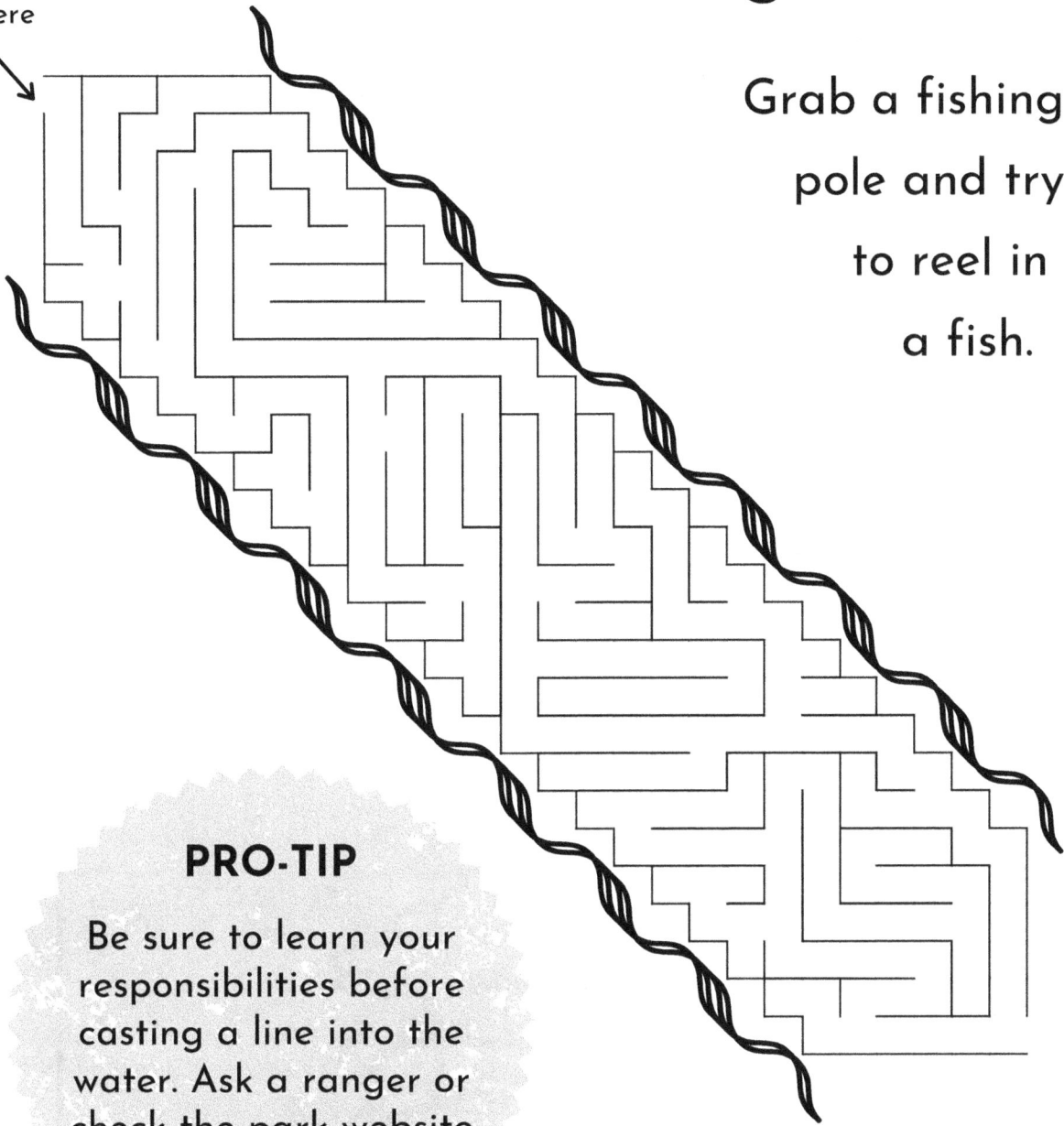

Grab a fishing
pole and try
to reel in
a fish.

PRO-TIP

Be sure to learn your
responsibilities before
casting a line into the
water. Ask a ranger or
check the park website
before you go.

Stacking Rocks

Have you ever seen stacks of rocks while hiking in national parks? Do you know what they are or what they mean? These rock piles are called cairns and often mark hiking routes in parks. Every park has a different way to maintain trails and cairns. However, they all have the same rule: If you come across a cairn, do not disturb it!

Color the cairn and the rules to remember.

1. Do not tamper with cairns.

If a cairn is tampered with or an unauthorized one is built, then future visitors may become disoriented or even lost.

2. Do not build unauthorized cairns.

Moving rocks disturbs the soil and makes the area more prone to erosion. Disturbing rocks can disturb fragile plants.

3. Do not add to existing cairns.

Authorized cairns are carefully designed. Adding to them can actually cause them to collapse.

Decoding Using American Sign Language

American Sign Language, also called ASL for short, is a language that many people who are deaf or hard of hearing use to communicate. People use ASL to communicate with their hands. Did you know people from all over the country and world travel to national parks? You may hear people speaking other languages. You might also see people using ASL. Use the American Manual Alphabet chart to decode some national parks facts.

This was the first national park to be established:

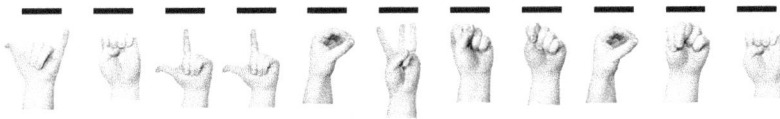

__ __ __ __ __ __ __ __ __ __

This is the biggest national park in the US:

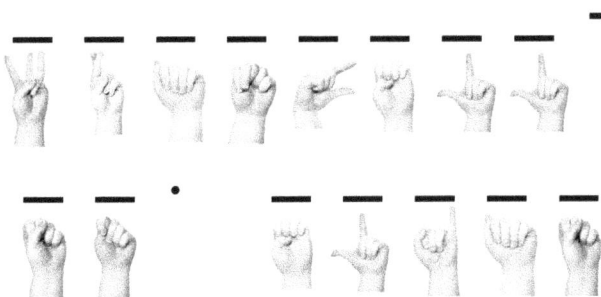

__ __ __ __ __ __ __ -

__ __ . __ __ __ __

This is the most visited national park:

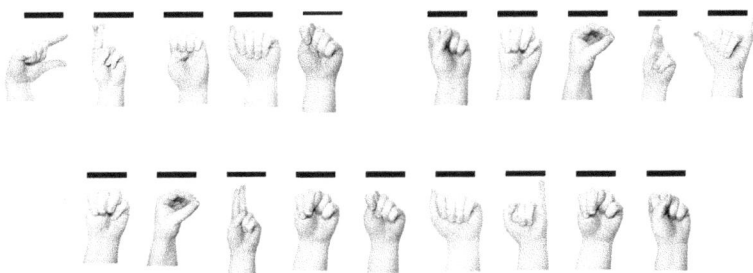

__ __ __ __ __ __ __ __

__ __ __ __ __ __

Aa	Bb	Cc	Dd	Ee
Ff	Gg		Hh	Ii
Jj	Kk	Ll	Mm	Nn
Oo	Pp	Qq		Rr
Ss	Tt	Uu		Vv
Ww	Xx	Yy	Zz	

Hint: Pay close attention to the position of the thumb!

Try it! Using the chart, try to make the letters of the alphabet with your hand. What is the hardest letter to make? Can you spell out your name? Show a friend or family member and have them watch you spell out the name of the national park you are in.

Go Horseback Riding to the Sequoia Grove

Help find the horse's lost shoe!

start here

DID YOU KNOW?

Horseback riding is a popular activity in Kings Canyon National Park. There are many trails that you can take horses for day trips.

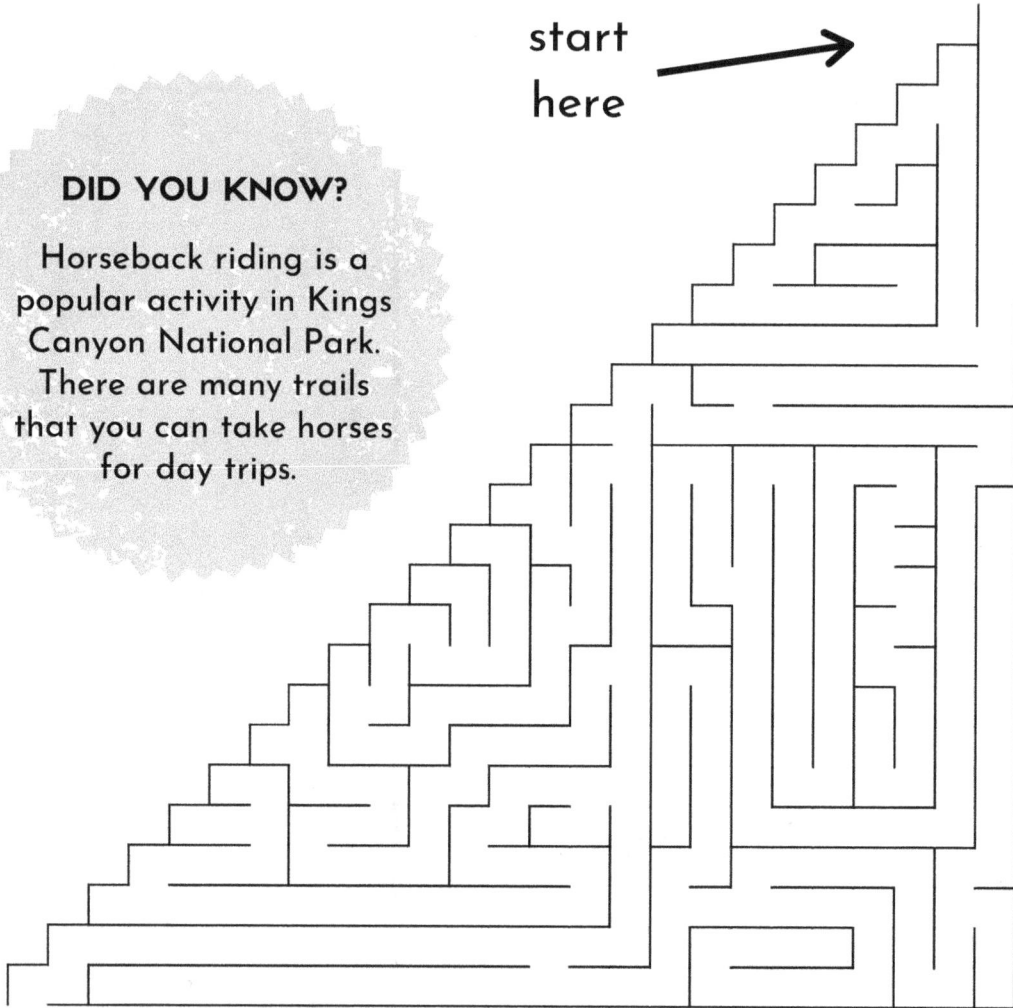

Butterflies of the Sierras

Dozens of species of butterflies and moths live in Kings Canyon National Park. Their wingspan size varies, as do the patterns on their wings. Design your own butterfly below. Make sure the wings are symmetrical, which means both sides match.

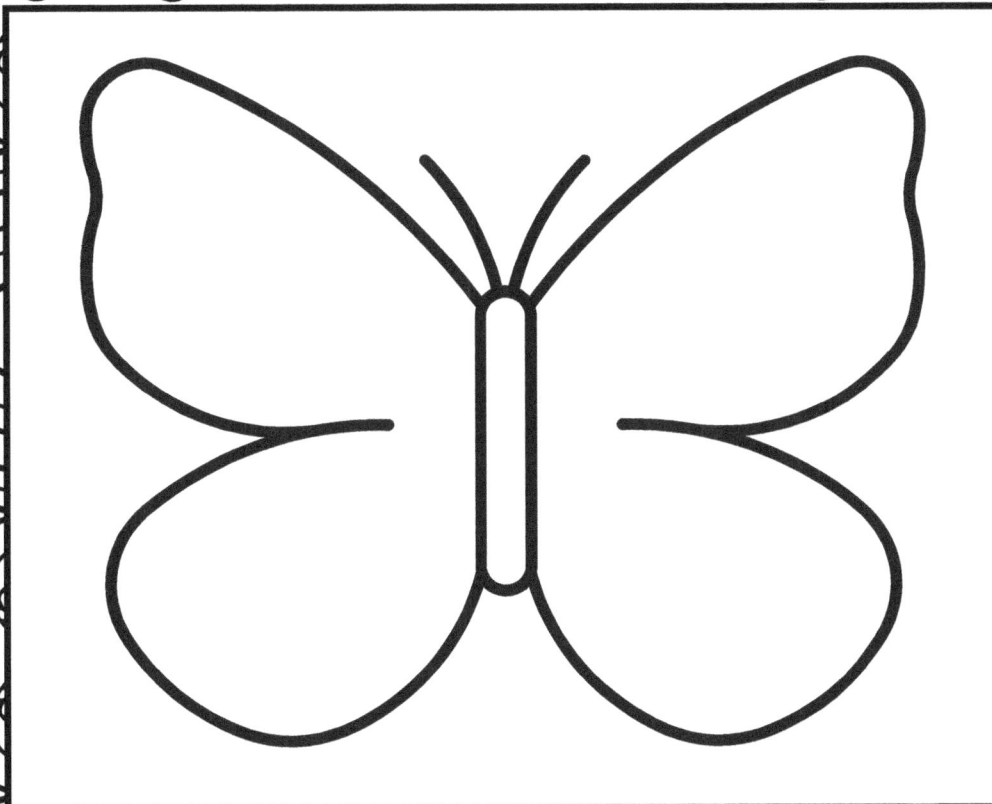

A Hike at Mist Falls

Fill in the blanks on this page without looking at the full story. Once you have each line filled out, use the words you've chosen to complete the story on the next page.

ADJECTIVE _____

SOMETHING TO EAT _____

SOMETHING TO DRINK _____

NOUN _____

ARTICLE OF CLOTHING _____

BODY PART _____

VERB _____

ANIMAL _____

SAME TYPE OF FOOD _____

ADJECTIVE _____

SAME ANIMAL _____

VERB THAT ENDS IN "ED" _____

NUMBER _____

A DIFFERENT NUMBER _____

SOMETHING THAT FLIES _____

LIGHT SOURCE _____

PLURAL NOUN _____

FAMILY MEMBER _____

YOUR NICKNAME _____

A Hike at Mist Falls

Use the words from the previous page to complete a silly story.

I went for a hike at Mist Falls today. In my favorite _ _ _ _ _ _ _ backpack, I
ADJECTIVE

made sure to pack a map so I wouldn't get lost. I also threw in an extra

_ _ _ _ _ _ _ _ _ _ just in case I got hungry and a bottle of _ _ _ _ _ _ _ _ _ _. I put
SOMETHING TO EAT   SOMETHING TO DRINK

on my _ _ _ _ _ _ _ _ _ spray, and I tied a _ _ _ _ _ _ _ _ _ _ _ around my
NOUN   ARTICLE OF CLOTHING

_ _ _ _ _ _ _ _ _ _, in case it gets chilly. I started to _ _ _ _ _ _ down the path. As
BODY PART   VERB

soon as I turned the corner, I came face to face with a(n) _ _ _ _ _ _ _ _. I think
ANIMAL

it was as startled as I was! What should I do? I had to think fast! Should I

give it some of my _ _ _ _ _ _ _ _ _ _ _? No. I had to remember what the
SAME TYPE OF FOOD

_ _ _ _ _ _ _ ranger told me: "If you see one, back away slowly and try not to
ADJECTIVE

scare it." Soon enough, the _ _ _ _ _ _ _ _ _ _ _ _ _ _ _ _ _ _ _ _ away. The coast
SAME ANIMAL   VERB THAT ENDS IN ED

was clear. _ _ _ _ _ _ hours later, I finally got to the lookout. I felt like I could
NUMBER

see for a _ _ _ _ _ _ miles. I took a picture of a _ _ _ _ _ _ _ _ so I could always
A DIFFERENT NUMBER   NOUN

remember this moment. As I was putting my camera away, a _ _ _ _ _ _ _ _ _
SOMETHING THAT FLIES

flew by, reminding me that it was almost nighttime. I turned on my

_ _ _ _ _ _ _ _ _ _ and headed back. I could hear the _ _ _ _ _ _ _ _ _ _ singing their
LIGHT SOURCE   PLURAL INSECT

evening song. Just as I was getting tired, I saw my _ _ _ _ _ _ _ _ _ _ and our tent.
FAMILY MEMBER

"Welcome back _ _ _ _ _ _ _! How was your hike?"
NICKNAME

Snail Mail

Design a postcard to send to a friend or family member. Who do you want to tell about Kings Canyon National Park? In the first template, write your message. In the second template, create a design for the front of the postcard. You could show something you saw, something you did, or something you want to do in the national park.

Postcard

Let's Go Camping Word Search

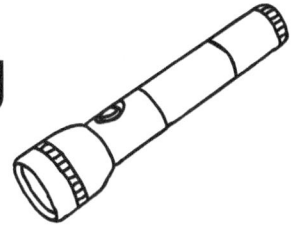

Words may be horizontal, vertical, diagonal, or they might even be backwards!

1. tent
2. camp stove
3. sleeping bag
4. bug spray
5. sunscreen
6. map
7. flashlight
8. pillow
9. lantern
10. ice
11. snacks
12. smores
13. water
14. first aid kit
15. chair
16. cards
17. books
18. games
19. trail
20. hat

```
D P P I L L O W D B T E A C I
E O A D P R E A A M B R C A N
P W C A M P S T O V E I H X G
R A H S G E L E B E E D A P S
E L B U G S P R A Y N G I E A
S I A H G C I C N N M E R C N
C W N L A F I R S K O O B F K
M T A E M I L E L H M R W L J
T A P R E A O R E S L B A A B
S M P A S R R T E N T L U S C
C E A I I R C G P E I U J H A
S S N A C K S S I M O K I L R
I J R S F O I S N J R A Q I D
C Y E T L E V E G U O R V G S
E W T A K C A B B S S O H H M
X J N F I R S T A I D K I T T
U A A E S S E N G E T P V A B
C J L I A R T D N A M A H A S
```

All in the Day of a Park Ranger

Park Rangers are hardworking individuals dedicated to protecting our parks, monuments, museums, and more. They take care of the natural and cultural resources for future generations. Rangers also help protect the visitors of the park. Their responsibilities are broad and they work both with the public and behind the scenes.

What have you seen park rangers do? Use your knowledge of the duties of park rangers to fill out a typical daily schedule, listing one activity for each hour. Feel free to make up your own, but some examples of activities are provided on the right. Read carefully! Not all the example activities are befitting a ranger.

6 am	Lead a sunrise hike
7 am	
8 am	
9 am	
10 am	
11 am	
12 pm	Enjoy a lunch break outside
1 pm	
2 pm	
3 pm	
4 pm	Teach visitors about the geology of the mountains
5 pm	
6 pm	
7 pm	
8 pm	
9 pm	

- feed the bald eagles
- build trails for visitors to enjoy
- throw rocks off the side of the mountain
- rescue lost hikers
- study animal behavior
- record air quality data
- answer questions at the visitor center
- pick wildflowers
- pick up litter
- share marshmallows with squirrels
- repair handrails
- lead a class on a field trip
- catch frogs and make them race
- lead people on educational hikes
- write articles for the park website
- protect the river from pollution
- remove non-native plants from the park
- study how climate change is affecting the park
- give a talk about mountain lions
- lead a program for campers on salmon

If you were a park ranger, which of the above tasks would you enjoy most?

_____

Draw Yourself as a Park Ranger

RANGER

The Giant Sequoias of the Sierra Nevadas

Uh oh! The names of these famous giant sequoias got mixed up! Unscramble the letters in each circle to figure out their names.

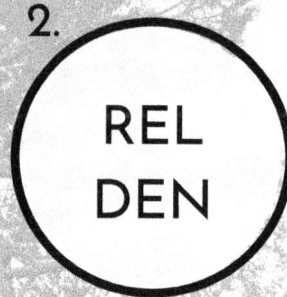

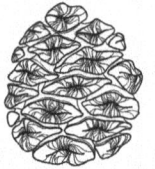

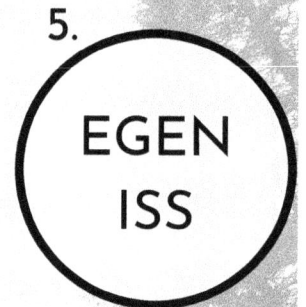

2.

REL DEN

1.

ERO NOM

5.

EGEN ISS

3.

LE BOO

4.

LRIN ANFK

1. _____

2. _____

3. _____

4. _____

5. _____

Word Bank

Monroe
Genesis
Boole
Stagg
Adam
Nelder
Hart
Franklin

Amphibians

Two species of toad and four species of frogs live in Kings Canyon National Park. Even more types of salamanders live there too. Frogs and toads both spend the beginning of their lives the same way - as tadpoles. Tadpoles hatch from eggs, usually in springs or pools of water.

Both frogs and toads are amphibians. Salamanders are amphibians too. Color the amphibians below.

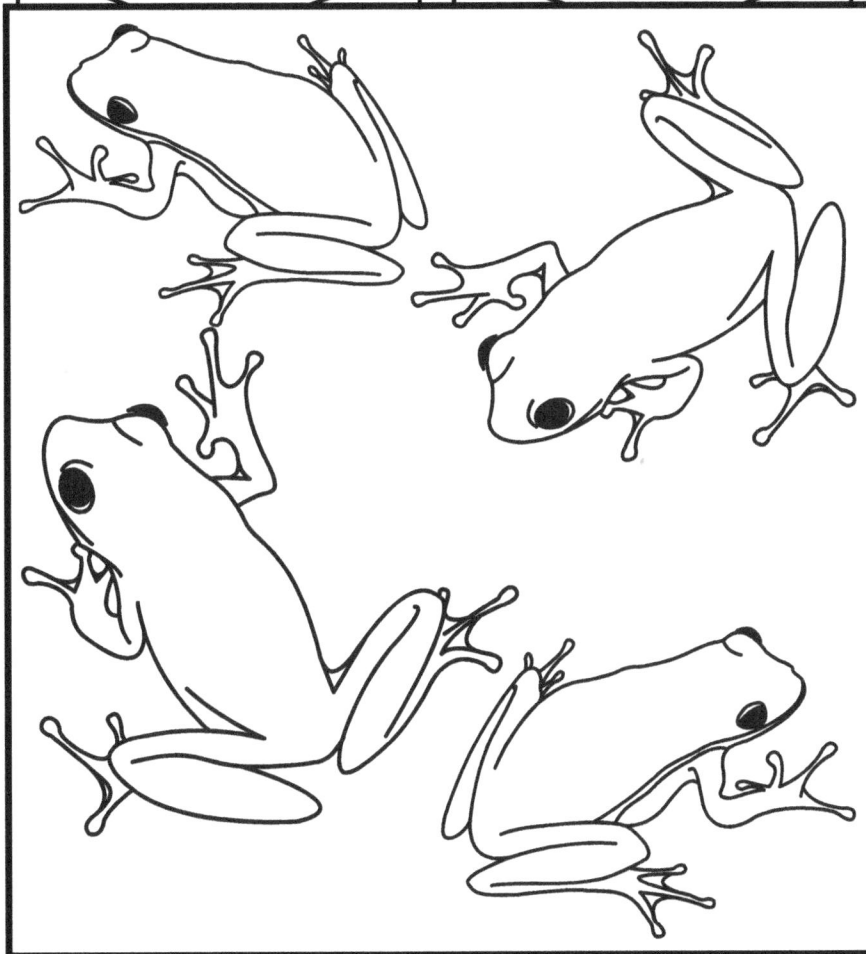

Being Respectful

Rangers need your help! Some people toss their trash where they shouldn't, create graffiti, or take artifacts when they visit Kings Canyon National Park. Create a poster to help show other visitors how to be respectful in the space below.

Rock Scavenger Hunt

Pay close attention to the things beneath your feet. If you visit Kings Canyon National Park, you will see all sorts of rocks, both big and small. Go on a rock hunt! You may have to get close to the ground and focus carefully to be able to find all the rocks on this list.

☐ A sharp rock

☐ A flat rock

☐ A round rock

☐ A rectangular rock

☐ A dull rock

☐ A rock with stripes

☐ A multicolored rock

☐ A smooth rock

☐ A small rock

☐ A huge rock

☐ A rough rock

☐ A shiny rock

☐ A rock with speckles

☐ A rock with only one color

Compare two rocks that look very different from each other.
What makes them different? Think about their size, their shape, their texture, and their color.
Do they have any similarities?

_____

_____

_____

63 National Parks

How many other national parks have you been to? Which one do you want to visit next? Note that if some of these parks fall on the border of more than one state, you may check it off more than once!

Alaska
☐ Denali National Park
☐ Gates of the Arctic National Park
☐ Glacier Bay National Park
☐ Katmai National Park
☐ Kenai Fjords National Park
☐ Kobuk Valley National Park
☐ Lake Clark National Park
☐ Wrangell-St. Elias National Park

American Samoa
☐ National Park of American Samoa

Arizona
☐ Grand Canyon National Park
☐ Petrified Forest National Park
☐ Saguaro National Park

Arkansas
☐ Hot Springs National Park

California
☐ Channel Islands National Park
☐ Death Valley National Park
☐ Joshua Tree National Park
☐ Kings Canyon National Park
☐ Lassen Volcanic National Park
☐ Pinnacles National Park
☐ Redwood National Park
☐ Sequoia National Park
☐ Yosemite National Park

Colorado
☐ Black Canyon of the Gunnison National Park
☐ Great Sand Dunes National Park
☐ Mesa Verde National Park
☐ Rocky Mountain National Park

Florida
☐ Biscayne National Park
☐ Dry Tortugas National Park
☐ Everglades National Park

Hawaii
☐ Haleakala National Park
☐ Hawai'i Volcanoes National Park

Idaho
☐ Yellowstone National Park

Kentucky
☐ Mammoth Cave National Park

Indiana
☐ Indiana Dunes National Park

Maine
☐ Acadia National Park

Michigan
☐ Isle Royale National Park

Minnesota
☐ Voyageurs National Park

Missouri
☐ Gateway Arch National Park

Montana
☐ Glacier National Park
☐ Yellowstone National Park

Nevada
☐ Death Valley National Park
☐ Great Basin National Park

New Mexico
☐ Carlsbad Caverns National Park
☐ White Sands National Park

North Dakota
☐ Theodore Roosevelt National Park

North Carolina
☐ Great Smoky Mountains National Park

Ohio
☐ Cuyahoga Valley National Park

Oregon
☐ Crater Lake National Park

South Carolina
☐ Congaree National Park

South Dakota
☐ Badlands National Park
☐ Wind Cave National Park

Tennessee
☐ Great Smoky Mountains National Park

Texas
☐ Big Bend National Park
☐ Guadalupe Mountains National Park

Utah
☐ Arches National Park
☐ Bryce Canyon National Park
☐ Canyonlands National Park
☐ Capitol Reef National Park
☐ Zion National Park

Virgin Islands
☐ Virgin Islands National Park

Virginia
☐ Shenandoah National Park

Washington
☐ Mount Rainier National Park
☐ North Cascades National Park
☐ Olympic National Park

West Virginia
☐ New River Gorge National Park

Wyoming
☐ Grand Teton National Park
☐ Yellowstone National Park

Other National Parks

Besides Kings Canyon National Park, there are 62 other diverse and beautiful national parks across the United States. Try your hand at this crossword. If you need help, look at the previous page for some hints.

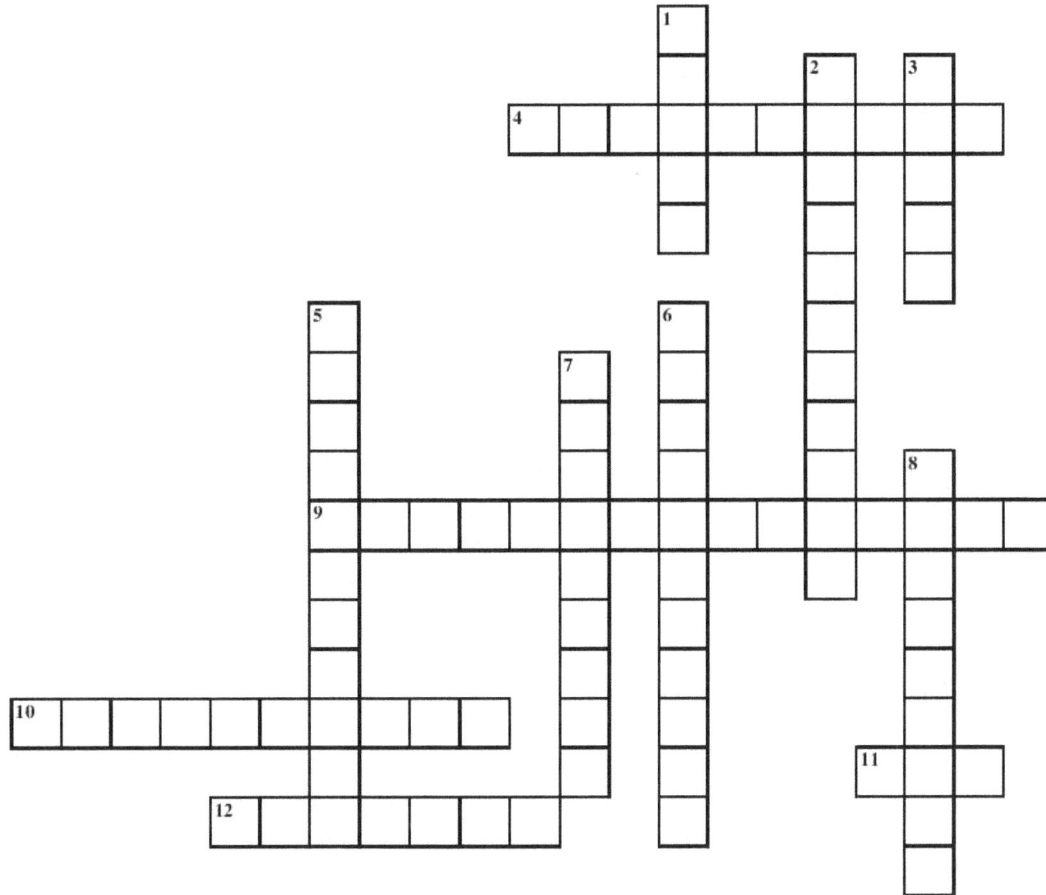

Down

1. State where Acadia National Park is located
2. This national park has the Spanish word for turtle in it
3. Number of national parks in Alaska
5. This national park has some of the hottest temperatures in the world
6. This national park is the only one in Idaho
7. This toothsome creature can famously be found in Everglades National Park
8. Only president with a national park named for them

Across

4. This state has the most national parks.
9. This park has some of the newest land in the US, caused by volcanic eruptions.
10. This park has the deepest lake in the United States.
11. This color shows up in the name of a national park in California.
12. This national park deserves a gold medal.

Which National Park Will You Go to Next?
Word Search

1. Zion
2. Big Bend
3. Glacier
4. Olympic
5. Sequoia
6. Bryce
7. Mesa Verde
8. Biscayne
9. Wind Cave
10. Great Basin
11. Katmai
12. Yellowstone
13. Voyageurs
14. Arches
15. Badlands
16. Denali
17. Glacier Bay
18. Hot Springs

```
F M M E S A V E R D E B N E Y
E A B I G B E N D E S A S E M
Y L I C A L O Y N E E D L T G
D M G A S S A U C N R L U E R
C E L I I T S C R E O A A K E
S N A W Y E E O I W T N A C A
G I C H A A Q C S E M D N S T
N O I Z P R U T I M R S N E B
I W E L M P O N B W E B K H A
R J R F D N I F L I H B U C S
P A B E E S A N E S O P W R I
S J A E N Y A C S I B A U A N
T C Y I A D O H H Y M E A L R
O T A T L M L E S E G R W R J
H S T O I K A T M A I R O P B
I C H U R C O L Y M P I C O U
O Y G T S D E O S B R Y C E T
W I N D C A V E I N R O H E M
```

Field Notes

Spend some time reflecting on your trip to Kings Canyon National Park. Your field notes will help you remember the things you experienced. Use the space below to write about your day.

While I was at Kings Canyon National Park...

I saw:

_____

_____

_____

_____

I heard:

_____

_____

_____

I felt:

_____

Draw a picture of your favorite thing in the park.

_____

_____

I wondered:

_____

_____

_____

_____

53

ANSWER KEY

National Park Emblem Answers

1. This represents all plants: **Sequoia Tree**

2. This represents all animals: **Bison**

3. This represents the landscapes: **Mountains**

4. This represents the waters protected by the park service: **Water**

5. This represents the historical and archeological values: **Arrowhead**

Jumbles Answers

1. ROCK CLIMBING

2. HIKING

3. BIRDING

4. CAMPING

5. PICNICKING

6. SIGHTSEEING

7. SNOWSHOEING

Go Birdwatching at Zumwalt Meadow

start here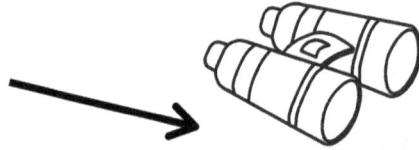

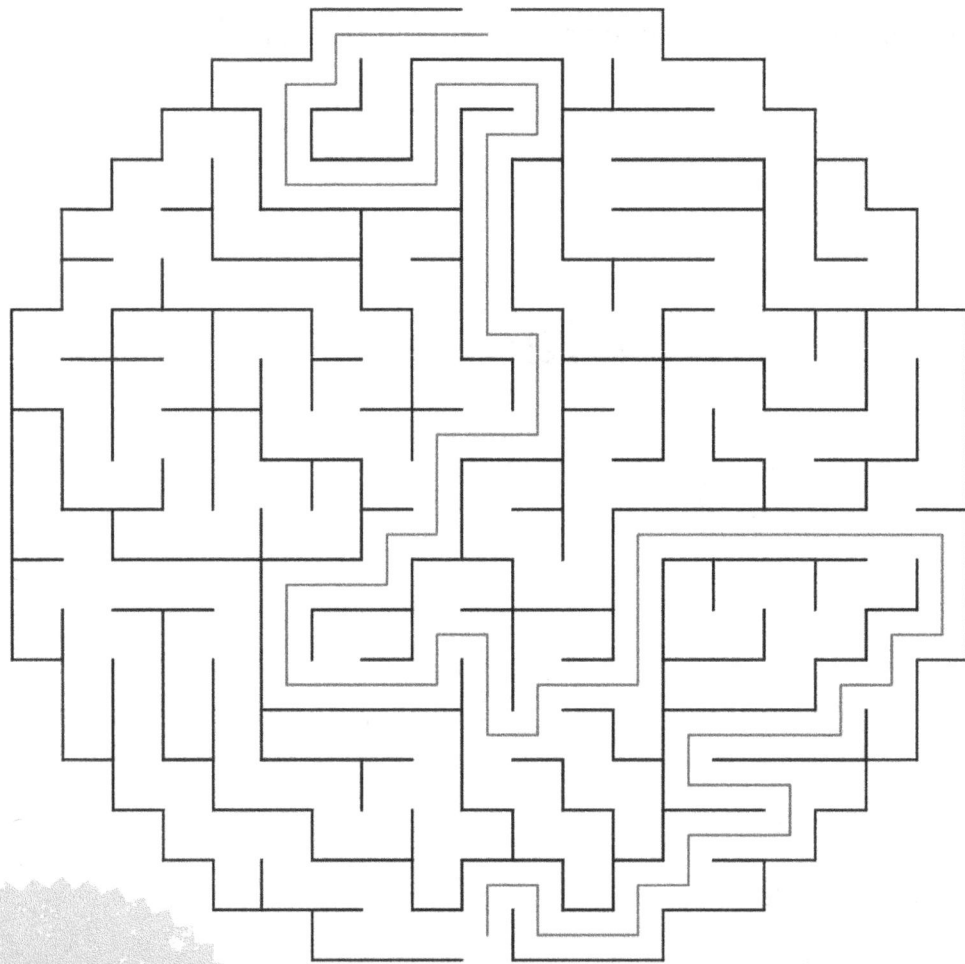

DID YOU KNOW?

Kings Canyon NP is home to several birds of prey, including eagles, hawks, and owls. Birds of prey are birds that hunt other animals for food.

Answers: Who Lives Here?

Below are 8 plants and animals that live in the park.
Use the word bank to fill in the clues below.

WORD BANK: GOPHER SNAKE, QUAIL, POSSUM, MARMOT, WESTERN TOAD, PIKA, BEAVER, WILD TURKEY

POS**S**UM

WEST**E**RN ▪ TOAD

QUAIL

WILD ▪ **T**URKEY

MARM**O**T

P**I**KA

BE**A**VER

GARTER ▪ **S**NAKE

Find the Match!
Common Names and Latin Names

Match the common name to the scientific name for each animal. The first one is done for you. Use clues on the page before and after this one to complete the matches.

Brewer's Blackbird

Giant Sequoia

Corn Lily

American Black Bear

Great Horned Owl

Bald Eagle

Ptarmigan

Pika

Rubber Boa

Haliaeetus leucocephalus

Ursus americanus

Lagopus leucura

Ochotona princeps

Sequoiadendron giganteum

Charina bottae

Bubo virginianus

Euphagus cyanocephalus

Veratrum californicum

Bald Eagle

Haliaeetus leucocephalus

Answers: The Ten Essentials

Careful preparation and knowledge are key to a successful adventure into Template's backcountry.

The ten essentials are a list of things that are important to have when you go for longer hikes. If you go on a hike to the <u>backcountry,</u> it is especially important that you have everything you need in case of an emergency. If you get lost or something unforeseen happens, it is good to be prepared to survive until help finds you.

The ten essentials list was developed in the 1930s by an outdoors group called the Mountaineers. Over time and technological advancements, this list has evolved. Can you identify all the things on the current list? Circle each of the "essentials" and cross out everything that doesn't make the cut.

(fire: matches, lighter, tinder, and/or stove)	~~a pint of milk~~	~~extra money~~	(headlamp, plus extra batteries)	(extra clothes)
(extra water)	~~a dog~~	~~Polaroid camera~~	~~bug net~~	~~lightweight game like a deck of cards~~
(extra food)	~~a roll of duct tape~~	(shelter)	(sun protection, such as sunglasses, sun-protective clothes and sunscreen)	(knife, plus a gear repair kit)
~~a mirror~~	(navigation: map, compass, altimeter, GPS device, or satellite messenger)	(first aid kit)	~~extra flip-flops~~	~~entertainment like video games or books~~

Backcountry - a remote undeveloped rural area.

Kings Canyon Word Search

Words may be horizontal, vertical, diagonal,
or they might be backwards!

1. giant sequoia
2. Fresno
3. trees
4. meadow
5. coyote
6. streams
7. California
8. cedar grove
9. pinecones
10. Panoramic Point
11. marble
12. Crystal Cave
13. pikas
14. marmot
15. skunk
16. forest
17. black bear

```
G W S L S P I R E F O R E S T
H I A S K I L S T R E A M S J
M E A D O W O S C E L B A P B
S P I N E C O N E S R L U A C
C E A D T A B L O N I U J N L
A T R E E S C O Y O T E A O I
L E E T E H E K I N K R I R N
I L V A M B I Q G W N E K A G
F E A S G L L O U E I D Y M M
O C C E D A R G R O V E O I A
R T L H C C I N O O I E M C N
N R A I K K E I S M O A I P E
I I T S H B I R E I R A L O W
A C S O L E V E S B S A K I P
N I Y K K A I N L R A L H N A
X T R F U R E E L Z E S Q T L
H Y C R O N L E C T R I C E E
F L O Y D N K D N T O M R A M
```

Answers: Find the Match!
What are Baby Animals Called?

Match the animal to its baby. The first one is done for you.

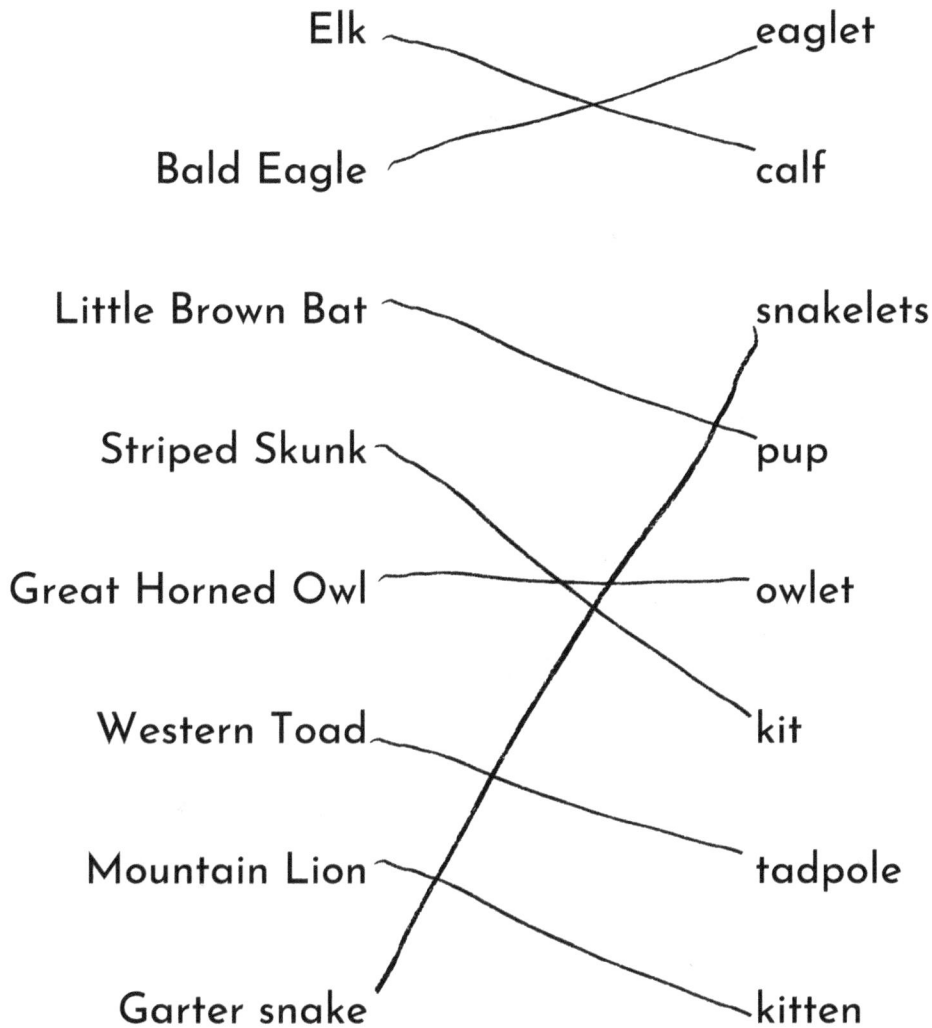

Elk — eaglet

Bald Eagle — calf

Little Brown Bat — snakelets

Striped Skunk — pup

Great Horned Owl — owlet

Western Toad — kit

Mountain Lion — tadpole

Garter snake — kitten

Hike to see the Giant Sequoias

DID YOU KNOW?

The General Grant tree in Kings Canyon National Park is the second largest tree in the world by volume.

The Biggest Trees
Word Search

Giant Sequoias are some of the biggest trees in the world. Many of the very largest trees are named and closely monitored. These are some of the biggest ones in Kings Canyon and Sequoia National Park. Can you find them?

1. General Sherman
2. General Grant
3. President
4. Lincoln
5. Stagg
6. Boole
7. Genesis
8. Franklin
9. Monroe
10. Column
11. Euclid
12. Pershing
13. Diamond
14. Adams
15. Nelder
16. Hart

```
G D E S C A N Y O N D E D W C
E E D P M I F R A N K L I N H
N V N R K I T E A W A L A O A
E E U E H A R T U Y U T M M T
S N N S R D Y P L C Y R O K R
I P D I Y A R E C T L E N O E
S O S D P M L R R H L I D A E
A R B E M S I S D I L S D N G
L T H N G I L H H N U D E C E
S I S T A G G I K E U G R O R
E S N U A E I N N L R B N L N
Q H N C K N O G S D S M T U E
U J O S O E I N Z E I O A M C
O Y G E L L V E I R D N V N O
I W E L D A N A D O A R H E M
A T G E N E R E N L B O O L E
U A E E S A E N N O A E V E B
C G E N E R A L G R A N T O N
```

Answers: Leave No Trace Quiz

Leave No Trace is a concept that helps people make decisions during outdoor recreation that protects the environment. There are seven principles that guide us when we spend time outdoors, whether you are in a national park or not. Are you an expert in Leave No Trace? Take this quiz and find out!

1. How can you plan ahead and prepare to ensure you have the best experience you can in the National Park?
    A. Make sure you stop by the ranger station for a map and to ask about current conditions.
2. What is an example of traveling on a durable surface?
    A. Walking only on the designated path.
3. Why should you dispose of waste properly?
    C. So that other peoples' experiences of the park are not impacted by you leaving your waste behind.
4. How can you best follow the concept "leave what you find?"
    B. Take pictures but leave any physical items where they are.
5. What is not a good example of minimizing campfire impacts?
    C. Building a new campfire ring in a location that has a better view.
6. What is a poor example of respecting wildlife?
    A. Building squirrel houses out of rocks from the river so the squirrels have a place to live.
7. How can you show consideration of other visitors?
    B. Wear headphones on the trail if you choose to listen to music.

All in the Day of a Park Ranger

There are many right answers for this activity, but not all of the provided examples are good activities for a park ranger. In fact, a park ranger's job may include stopping visitors from doing some of these things.

The list below has activities that rangers do not do:

feed the migratory birds

throw rocks off the side of the mountain

pick wildflowers

share marshmallows with squirrels

catch frogs or toads and make them race

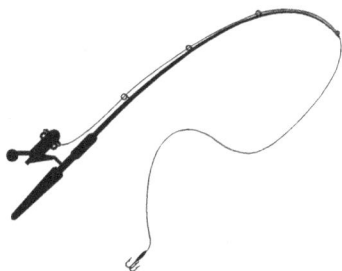

Solution: Catch a Fish in the Kings River

Grab a fishing pole and try to reel in a fish.

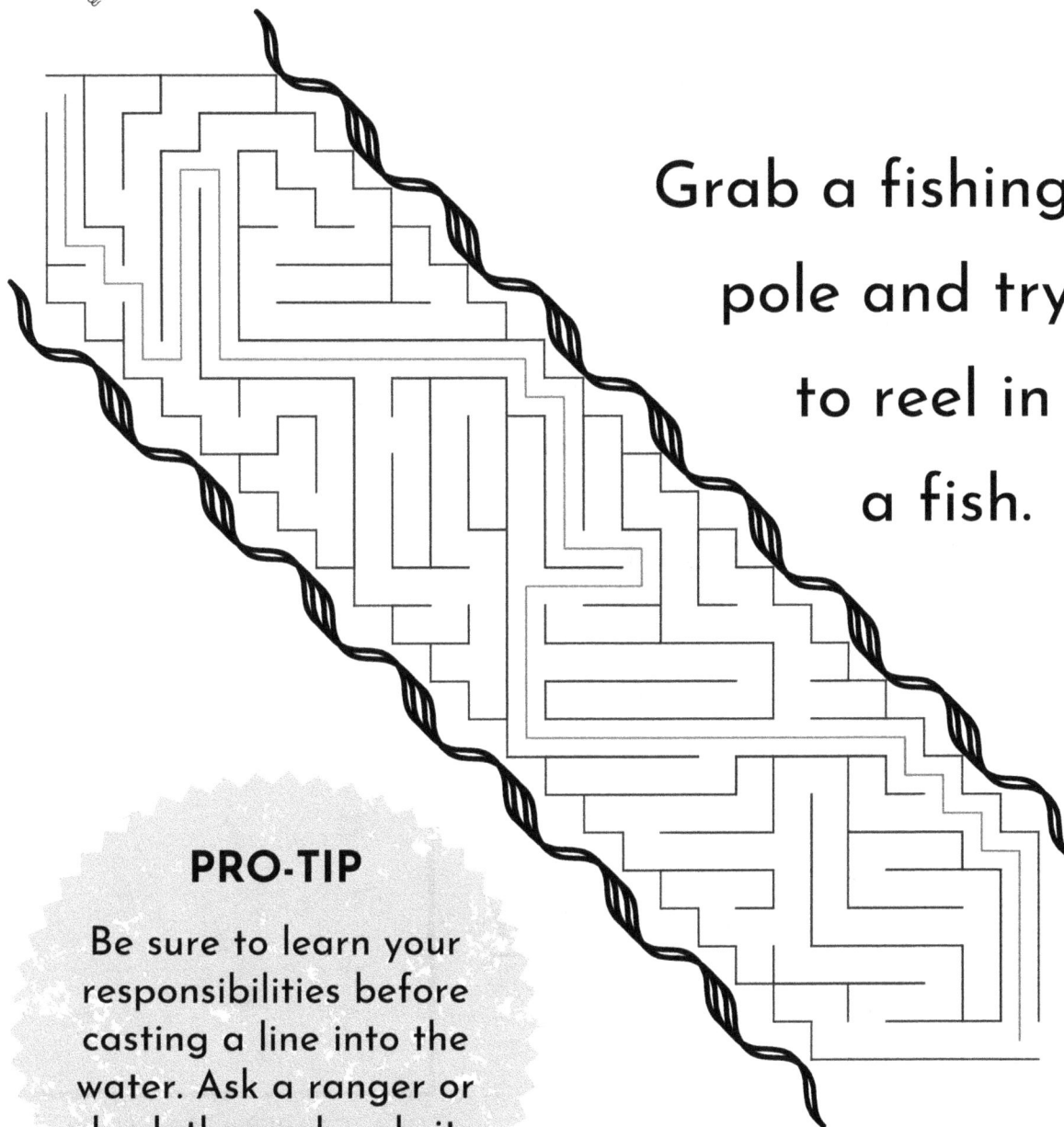

PRO-TIP

Be sure to learn your responsibilities before casting a line into the water. Ask a ranger or check the park website before you go.

Decoding Using American Sign Language

American Sign Language, also called ASL for short, is a language that many people who are deaf or hard of hearing use to communicate. People use ASL to communicate with their hands. Did you know people from all over the country and world travel to national parks? You may hear people speaking other languages. You might also see people using ASL. Use the American Manual Alphabet chart to decode some national parks facts.

This was the first national park to be established:

Y E L L O W S T O N E

This is the biggest national park in the US:

W R A N G E L L -

S T .   E L I A S

This is the most visited national park:

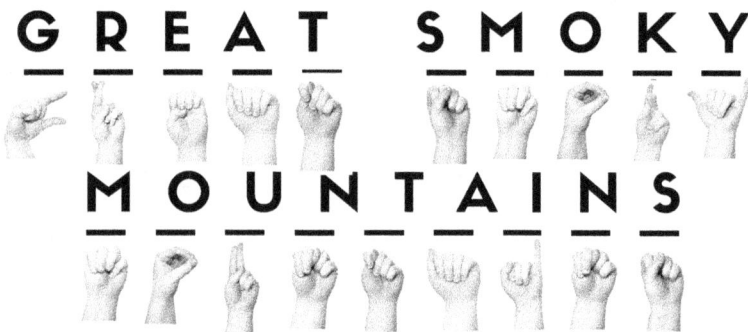

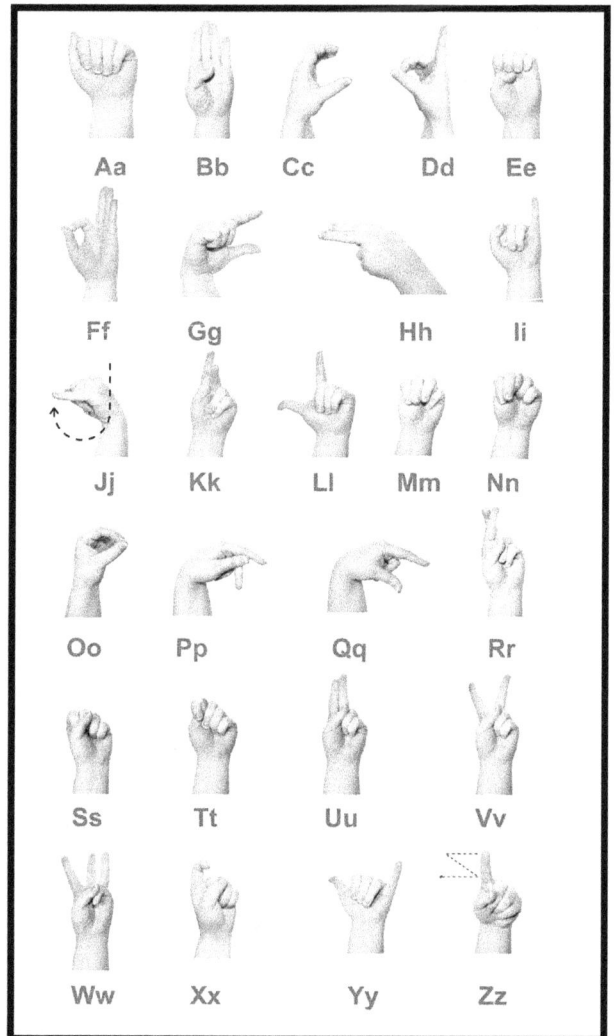

G R E A T   S M O K Y

M O U N T A I N S

Hint: Pay close attention to the position of the thumb!

Try it! Using the chart, try to make the letters of the alphabet with your hand. What is the hardest letter to make? Can you spell out your name? Show a friend or family member and have them watch you spell out the name of the national park you are in.

Go Horseback Riding on the Sequoia Grove

Help find the horse's lost shoe!

start here →

DID YOU KNOW?

Horseback riding is a popular activity in Kings Canyon National Park. There are many trails that you can take horses for day trips.

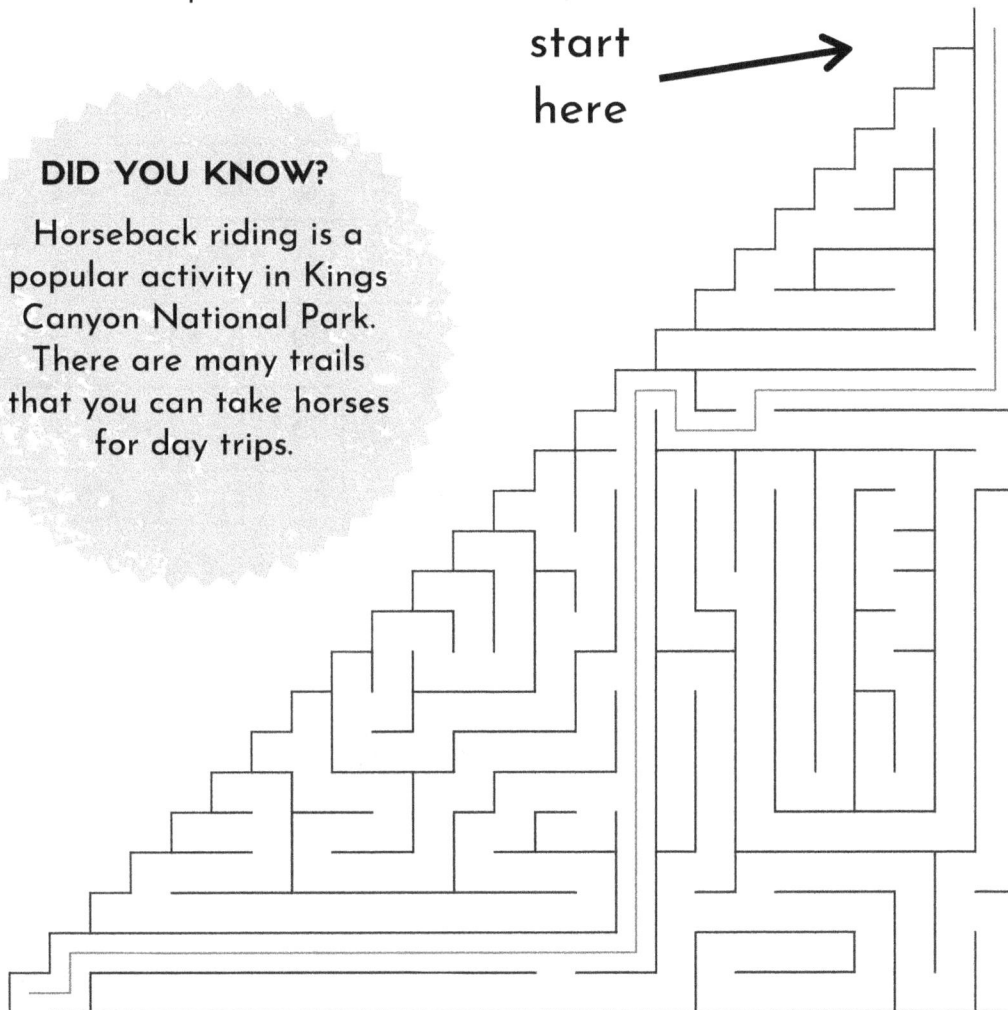

Let's Go Camping
Word Search

1. tent
2. camp stove
3. sleeping bag
4. bug spray
5. sunscreen
6. map
7. flashlight
8. pillow
9. lantern
10. ice
11. snacks
12. smores
13. water
14. first aid kit
15. chair
16. cards
17. books
18. games
19. trail
20. hat

```
D P P I L L O W D B T E A C I
E O A D P R E A A M B R C A N
P W C A M P S T O V E I H X G
R A H S G E L E B E E D A P S
E L B U G S P R A Y N G I E A
S I A H G C I C N N M E R C N
C W N L A F I R S K O O B F K
M T A E M I L E L H M R W L J
T A P R E A O R E S L B A A B
S M P A S R R T E N T L U S C
C E A I I R C G P E I U J H A
S S N A C K S S I M O K I L R
I J R S F O I S N J R A Q I D
C Y E T L E V E G U O R V G S
E W T A K C A B B S S O H H M
X J N F I R S T A I D K I T T
U A A E S S E N G E T P V A B
C J L I A R T D N A M A H A S
```

The Giant Sequoias of the Sierra Nevadas

Uh oh! The names of these famous giant sequoias got mixed up! Unscramble the letters in each circle to figure out their names.

2. REL DEN

ERO NOM

5. EGEN ISS

3. LE BOO

4. LRIN ANFK

1. <u>MONROE</u>
2. <u>NELDER</u>
3. <u>BOOLE</u>
4. <u>FRANKLIN</u>
5. <u>GENESIS</u>

Word Bank

Monroe
Genesis
Boole
Stagg
Adam
Nelder
Hart
Franklin

Answers: Other National Parks

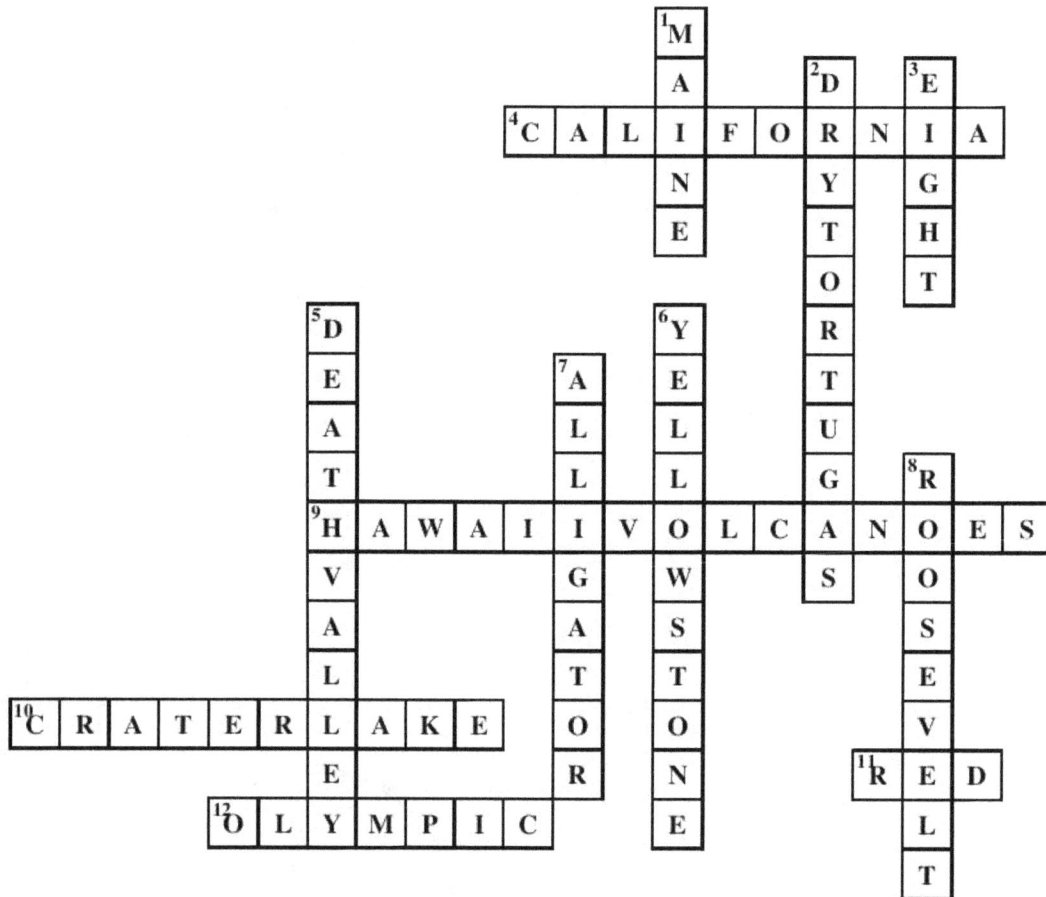

```
              ¹M
          A       ²D      ³E
    ⁴C A L I F O R N I A
          N       Y       G
          E       T       H
                  O       T
    ⁵D            R
    E       ⁶Y    T
    A   ⁷A  E    U
    T   L   L    G    ⁸R
    ⁹H A W A I I V O L C A N O E S
    V   G   W    S    O
    A   A   S         S
    L   T   T         E
    ⁰C R A T E R L A K E   V
    E   O   O         ¹R E D
    ¹²O L Y M P I C   N     L
        R   E               T
```

Down

1. State where Acadia National Park is located
2. This National Park has the Spanish word for turtle in it
3. Number of National Parks in Alaska
5. This National Park has some of the hottest temperatures in the world
6. This National Park is the only one in Idaho
7. This toothsome creature can famously be found in Everglades National Park
8. Only president with a national park named for them

Across

4. This state has the most National Parks
9. This park has some of the newest land in the US, caused by a volcanic eruption
10. This park has the deepest lake in the United States
11. This color shows up in the name of a National Park in California
12. This National Park deserves a gold medal

Answers: Where National Park Will You Go Next?

1. Zion
2. Big Bend
3. Glacier
4. Olympic
5. Sequoia
6. Bryce
7. Mesa Verde
8. Biscayne
9. Wind Cave
10. Great Basin
11. Katmai
12. Yellowstone
13. Voyageurs
14. Arches
15. Badlands
16. Denali
17. Glacier Bay
18. Hot Springs

```
F M M E S A V E R D E B N E Y
E A B I G B E N D E S A S E M
Y L I C A L O Y N E E D L T G
D M G A S S A U C N R L U E R
C E L I I T S C R E O A A K E
S N A W Y E E O I W T N A C A
G I C H A A Q C S E M D N S T
N O I Z P R U T I M R S N E B
I W E L M P O N B W E B K H A
R J R F D N I F L I H B U C S
P A B E E S A N E S O P W R I
S J A E N Y A C S I B A U A N
T C Y I A D O H H Y M E A L R
O T A T L M L E S E G R W R J
H S T O I K A T M A I R O P B
I C H U R C O L Y M P I C O U
O Y G T S D E O S B R Y C E T
W I N D C A V E I N R O H E M
```

LITTLE BISON

Press

Little Bison Press is an independent children's book publisher based in the Pacific Northwest. We promote exploration, conservation, and adventure through our books. Established in 2021, our passion for outside spaces and travel inspired the creation of Little Bison Press.

We seek to publish books that support children in learning about and caring for the natural places in our world.

To learn more, visit:
www.littlebisonpress.com

Want more free games and activities? Visit our website!